Congressional
Research
Service

# U.S. Renewable Electricity: How Does Wind Generation Impact Competitive Power Markets?

Phillip Brown
Specialist in Energy Policy

November 7, 2012

Congressional Research Service

7-5700

www.crs.gov

R42818

# Summary

U.S. wind power generation has experienced rapid growth in the last 20 years as total installed capacity has increased from 1,500 megawatts (MW) in 1992 to more than 50,000 MW in August of 2012. According to the Energy Information Administration (EIA), wind power provided approximately 3% of total U.S. electricity generation in 2011. Two primary policies provide market and financial incentives that support the wind industry and have contributed to U.S. wind power growth: (1) production tax credit (PTC)—a federal tax incentive of 2.2 cents for each kilowatt-hour (kWh) of electricity produced by a qualified wind project (set to expire for new projects at the end of 2012), and (2) renewable portfolio standards (RPS)—state-level policies that encourage renewable power by requiring that either a certain percentage of electricity be generated by renewable energy sources or a certain amount of qualified renewable electricity capacity be installed.

The concentration of wind power projects within competitive power markets managed by regional transmission operators (RTOs), the focus of this report, has resulted in several concerns expressed by power generators and other market participants. Three specific concerns explored in this report include: (1) How might wind power affect wholesale market clearing prices? (2) Does wind power contribute to negative wholesale power price events? and (3) Does wind power impact electric system reliability? These concerns might be considered during congressional debate about the future of wind PTC incentives.

When considering the potential impacts of wind power on electric power markets, it is important to recognize that wholesale power markets are both complex and multi-dimensional. Wholesale power markets are influenced by a number of factors, including weather, electricity demand, natural gas prices, transmission constraints, and location. Therefore, determining the direct impact of a single variable, in this case wind power, on the financial economics of power generators can be difficult. In 2012, wholesale electric power prices were down from recent highs in 2008, and lower price trends can result in financial pressure for power generators in RTO markets. Arguably, however, the two primary contributors to this decline are low natural gas prices and low electricity demand.

Wind power generation can potentially reduce wholesale electricity prices, in certain locations and during certain seasons and times of day, since wind typically bids a zero ($0.00) price into wholesale power markets. Additionally, independent market monitor reports for three different RTOs each indicate that wind generators will sometimes bid a negative wholesale price in order to ensure electricity dispatch. The ability of wind generators to bid negatively priced power is generally attributed to value associated with PTC incentives and the ability to sell renewable energy credits (REC). However, wholesale power price reductions and negative electricity prices associated with wind generation need to be considered in context with other dimensions of organized power markets. For example, other revenue sources (i.e., capacity markets) may be available to generators in certain RTO market areas. Also, generators oftentimes enter into bilateral power purchase agreements that can provide a hedge against power price volatility. Therefore, the absolute impact of wind electricity on the economics of power generators is difficult to determine due to the many variables and dimensions that influence wholesale power markets.

With regard to how wind power might impact electricity system reliability, two aspects of reliability are typically discussed: (1) impacts to system operations—the ability of the power system to manage the variable and sometimes unpredictable nature of wind power production, and (2) resource adequacy and capacity margins—the potential for wind power generation to either influence power plant retirements or contribute to market conditions that do not support investment in new capacity resources. RTOs are currently implementing various initiatives (i.e., dispatchable resource programs, renewable energy transmission projects) to address the variable generation characteristics of wind power. Furthermore, each RTO market is designed to provide the economic signals necessary to stimulate capacity additions in order to ensure resource adequacy and maintain capacity margins. However, should wind power generation continue to grow, it is uncertain if current RTO market designs will provide the signals needed to encourage specific types of generation capacity (e.g., operating and spinning reserves) necessary to manage the variable nature of wind power.

# Contents

# Figures

## Tables

## Contacts

# Overview

This report analyzes the impacts of wind generation on competitive power markets, including financial and economic impacts on electric power generators. Overall, the goal of this report is to provide context for several electricity market concepts that are relevant to understanding the economic effects of wind power generation. Additionally, this report addresses three specific questions about the market interaction of wind power and electric power generators: (1) How might wind power affect wholesale market clearing prices? (2) Does wind power contribute to negative wholesale power price events within competitive electric power markets? and (3) Does wind power impact electric system reliability? This report focuses on data and information available for competitive electricity markets that are managed by a regional transmission operator (RTO) or independent system operator (ISO). Specific information for three RTO/ISO organizations is provided in this report: (1) Midwest Independent System Operator (MISO), (2) PJM, and (3) Electric Reliability Council of Texas (ERCOT). These three RTOs were selected for the analysis in an effort to limit the scope of this report. Furthermore, these RTOs are commonly cited as markets that are being affected by wind power generation. As a result, there is no discussion of wind power market impacts within cost-of-service, vertically integrated electricity markets that are common in the West and Southeast regions of the United States, nor is there any discussion of how wind power is managed by federally owned transmission system operators such as the Bonneville Power Administration.

# Background

Wind electricity generation in the United States has experienced rapid growth over the last two decades. In 1992, cumulative installed U.S. wind capacity was approximately 1,500 megawatts (MW).[1] In August of 2012, the U.S. wind industry reached a new milestone of 50,000 MW of installed wind power capacity.[2] In 2011, 3% of electricity generated in the United States was derived from wind power, compared to less than 1% in 1992.[3]

Growth in the U.S. wind power sector has been influenced by complementary policies and incentives at both the state and federal level. States essentially create demand for wind power projects by implementing renewable portfolio standard (RPS) policies that require a certain amount of renewable power to be generated by a certain date. For example, a state-level RPS may require that 25% of retail electricity sales be derived from renewable energy sources by 2025. As of September 2012, 29 states and the District of Columbia had established binding RPS policies.[4] Each state RPS policy is unique with respect to its design, goals, and means of compliance.

Federal incentives that support wind electricity generation are generally provided in the form of tax credits and depreciation benefits for qualified wind power projects. Federal tax incentives for

---

[1] Wiser, R, and Bollinger, M., "2008 Wind Technologies Market Report," U.S. Department of Energy, July 2009, available at http://www.windpoweringamerica.gov/pdfs/2008_annual_wind_market_report.pdf.

[2] "American Wind Power Reaches 50-gigawatt Milestone," American Wind Energy Association, August 7, 2012.

[3] U.S. Energy Information Administration, http://www.eia.gov/energyexplained/index.cfm?page= electricity_in_the_united_states.

[4] Database of State Incentives for Renewables and Efficiency, available at http://www.dsireusa.org/documents/ summarymaps/RPS_map.pdf.

---

wind power provide an economic incentive for wind projects and they may reduce the financial impact to rate payers in states that require a certain amount of renewable electricity generation. Currently, qualified wind power projects can receive either a production tax credit for every kilowatt-hour of electricity produced during the first 10 years of operation or a project can elect to receive a one-time investment tax credit equal to 30% of qualified project capital costs. The federal production tax credit is currently 2.2 cents ($0.022) for each kilowatt-hour of electricity produced from a qualified wind project. Under current law, tax credit incentives for new wind power projects are scheduled to end on January 1, 2013.[5]

U.S. wind power growth has resulted in several challenges that are being addressed in various power markets throughout the country. The variable nature of wind power generation has led to several integration studies aimed at analyzing transmission system operational challenges that may need to be addressed as wind power continues to grow.[6] Additionally, as will be discussed further in this report, some U.S. electricity markets have experienced lower, and even negative, wholesale market prices at certain locations at certain times that may be linked to wind power generation. Reports by consulting companies have been published indicating that low and negative wholesale prices resulting from wind power are distorting markets, adversely impacting the financial economics of conventional power generators, and skewing the economic signals necessary to stimulate new capacity builds to ensure electric system reliability.[7] However, other studies have concluded that additional wind power production within RTO markets will reduce wholesale prices and ultimately reduce electricity costs to consumers.[8]

Competitive electricity markets are extremely complex and vary regionally, with different markets having different rules regarding electricity dispatch, and revenue opportunities, among others. Additionally, the financial performance of electricity generators operating in competitive power markets is multi-dimensional and is influenced by a number of variables (e.g., natural gas costs, coal costs, electricity demand, and transmission congestion). As a result, identifying and quantifying the direct and absolute impact of one factor, such as wind power generation, on competitive markets can be difficult.

A comprehensive analysis of electricity markets and how wind power affects each respective market is beyond the scope of this report. Generally, however, there are two distinct types of markets in the United States: (1) competitive markets, in which power generators are subject to price competition when selling power into wholesale markets, and (2) cost-of-service markets, in which power generators earn a regulated rate-of-return established by a public utilities commission.[9] This report's analysis is confined to how wind power is integrated into competitive markets.

---

[5] For additional information about how the federal PTC impacts the U.S. wind industry, see CRS Report R42576, *U.S. Renewable Electricity: How Does the Production Tax Credit (PTC) Impact Wind Markets?*, by Phillip Brown.

[6] One specific study that evaluates wind integration issues is "Eastern Wind Integration and Transmission Study," National Renewable Energy Laboratory, January 2010, available at http://www.nrel.gov/wind/systemsintegration/ewits.html.

[7] F. Huntowski, A. Patterson, and M. Schnitzer, "Negative Electricity Prices and the Production Tax Credit," The Northbridge Group, September 14, 2012.

[8] B. Fagan, et al., "The Potential Rate Effects of Wind Energy and Transmission in the Midwest ISO Region," Synapse Energy Economics, Inc., May 22, 2012.

[9] For additional background on the U.S. power sector, see "Energy Primer: A Handbook of Energy Market Basics," Federal Energy Regulatory Commission, July 2012.

---

# Key Concepts and Definitions

In order to provide some context for the discussions below about how wind power might affect wholesale electric power markets, an overview of some important concepts and definitions is provided to orient the reader. These concepts and definitions provide some insight into the complexities associated with U.S. power markets, and the relationships between wind power generation, wholesale power markets, and the financial economics of power generators.

## Regional Transmission Operators /Independent System Operators

RTO/ISO organizations are generally wholesale power markets operated by an independent third party entity. The RTO/ISO controls the wholesale power and transmission system within a defined area and is responsible, in part, for balancing electricity supply with demand, and dispatching generation, through market price mechanisms. There are seven RTO/ISO organizations in the United States (see **Figure 1**) that serve approximately two-thirds of U.S. electric power customers.[10] Each RTO/ISO organization has a different structure and set of market rules that members must follow in order to participate in each respective market. How wind power might affect other generators depends on the structure of specific markets, and rules for dispatch of power plants.

**Figure 1. RTO/ISO Markets in the Contiguous United States**

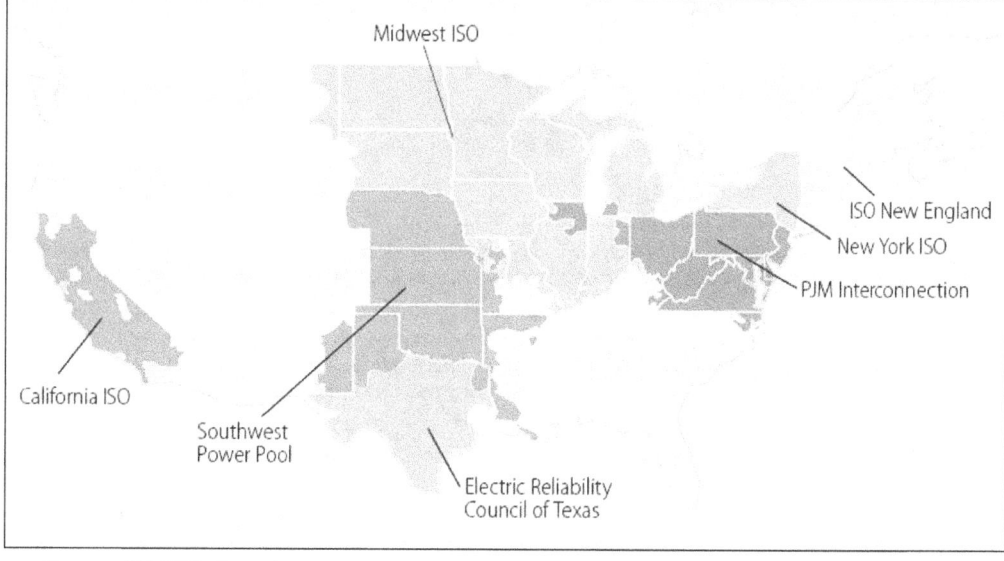

**Source:** ISO/RTO Council.

**Notes:** The Midwest ISO reliability coordination area includes Manitoba, Canada. Alaska and Hawaii, not shown in the figure, are not RTO/ISO markets.

---

[10] ISO/RTO Council, http://www.isorto.org, accessed September 10, 2012.

## Economic Supply and Electricity Dispatch

In order to supply reliable power at the lowest possible cost, wholesale electricity is typically dispatched to demand centers based on the lowest marginal cost of electricity available to satisfy demand. Generally, the marginal cost of electric power is based in part on the cost of fuel needed to generate a unit of electricity (typically measured in megawatt-hours) as well as the efficiency of the power plant. In competitive markets, the RTO/ISO will request wholesale price bids from all generators within the respective system for a defined period of time. Once bids are received, the ISO typically arranges the bids from lowest to highest in order to create what is known as a "dispatch curve." **Figure 2** provides an example illustration of how a dispatch curve might look.

**Figure 2. Simplified Electricity Dispatch Curve for Wholesale Power**

(Hypothetical RTO market example for a one-hour period)

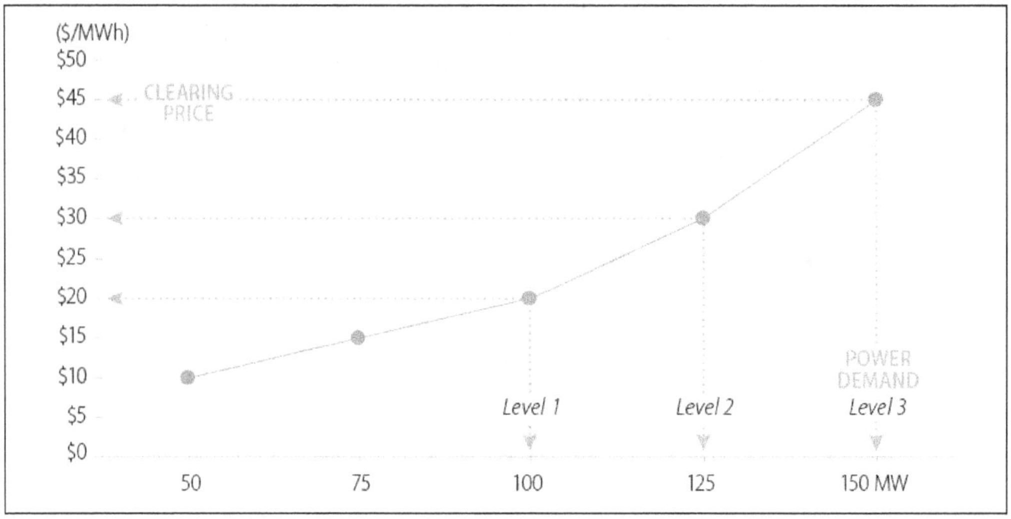

**Source:** CRS adaptation of *Locational Marginal Pricing (LMP) Overview*, page 22, PJM, October 18, 2011, available at https://www.pjm.com/training/~/media/training/core-curriculum/ip-lse-201/lmp-overview.ashx.

**Notes:** This figure illustrates how the wholesale electricity clearing price might change as a function of power demand. The red line represents the bid offer prices for electricity that are organized from low to high. Depending on the level of demand (three different hypothetical levels illustrated in this figure), the clearing price is adjusted in order to satisfy required demand for electricity delivery during a certain time period.

MW = megawatts

MWh = megawatt-hours

The slope of the dispatch curve and the clearing price can differ in each location based on factors such as the types of generators providing bids into the market, the season and time-of-day, transmission congestion, and the demand level served during a particular interval.

## Transmission Congestion Constraints

Transmission congestion can occur when the amount of power attempting to use a certain transmission corridor exceeds the transfer capacity of that transmission infrastructure. For example, if a transmission line has a transfer capacity of 300 MW and the combination of generation resources attempting to use that transmission line during a certain period is 400 MW, then the RTO must manage generation in order to not exceed the transfer limitations of the line. Economic dispatch depends on an electric system having sufficient transmission capability to deliver power from the chosen generating units. When this is not the case, system operators will lower the output of lower price units in one area and raise the output of higher cost units in another area until the expected loading of the transmission lines between the two areas is equal to their carrying capacity. Another way of thinking about this is that when transmission constraints result from the initial dispatch result, the model will re-dispatch the two regions separately. This yields different prices in the two areas. Transmission congestion is typically cited as one of the primary causes of negative prices in RTO/ISO markets.

## Locational Marginal Prices (LMP)

RTO/ISO electricity markets use locational marginal prices (LMP) in order to reflect individual prices that are determined for different zones and/or nodes within the RTO's system. Zones/nodes are typically located near power generators and at load busses where power is delivered to load-serving entities' distribution systems. LMPs are marginal because they are set based on the bid of the first generator not taken in the supply stack.[11] While RTOs produce prices at many locations, the prices are not always different. However, differences can arise due to transmission congestion, as discussed above. Instead of having a single wholesale price across the entire RTO, different prices are set in different areas in order to manage transmission constraints that may exist within the respective power market. Generators in areas with higher LMPs will be paid a higher price and generators in areas with lower LMPs will be paid a lower price.

## Multiple Revenue Sources

Power generators operating in competitive markets may have multiple revenue opportunities. Generally, these opportunities can include: (1) energy sales—the units of electricity (megawatt-hours) sold into the wholesale market, (2) capacity payments—payments made to generators in order to maintain resource adequacy for reliability requirements, and (3) ancillary services—including services such as frequency regulation and voltage control. Energy sales typically represent the largest revenue opportunity for generators in competitive markets. As briefly discussed above, each wholesale power market is different and revenue opportunities can vary by market and region. For example, PJM, the operator in the mid-Atlantic region, has a capacity market and an energy market in which generators participate. However, ERCOT only has an energy market. Material presented and discussed in this report is primarily focused on energy sales impacts—impacts of other revenue sources are not discussed. **Table 1** compares the generation capacity, wind capacity, and market types for MISO, PJM, and ERCOT.

---

[11] For more information about locational marginal prices, see http://www.pjm.com/sitecore/content/Globals/Training/Courses/ol-lmp-ov.aspx.

**Table 1. Overview Comparison of Generation Capacity, Wind Capacity, and Market Types**

| RTO/ISO | Total Generation Capacity (GW) | Wind Capacity (GW) | Markets | | | |
|---------|---------|---------|-----------|-----------|-----------|----------|
| | | | Day Ahead | Real Time | Anc. Serv. | Capacity |
| MISO | 132 | 12 | Yes | Yes | Yes | Yes (voluntary) |
| PJM | 186 | 5 | Yes | Yes | Yes (regulation, synchronous reserves, scheduled reserves) | Yes (mandatory) |
| ERCOT | 84 | 11 | Yes | Yes | Yes (included in day-ahead) | No |

**Source:** PJM, MISO, and ERCOT web sites and market monitor reports, "Energy Primer: A Handbook of Energy Market Basics," Federal Energy Regulatory Commission, July 2012.

**Notes:** Day Ahead and Real Time markets are energy markets.

Anc. Serv. = Ancillary Services

As indicated in **Table 1**, each RTO has a different level of wind penetration and has different types of markets that are available to all participating generators. The most notable differences are with capacity markets; for example, ERCOT does not operate a capacity market, while MISO has a voluntary capacity market.

## Day-Ahead and Real-Time Prices

As described in **Table 1**, RTOs manage several markets as part of their operations. With regard to energy sales, the most commonly monitored prices are those for day-ahead and real-time energy sales. For each hour of each following day, RTOs will set market clearing prices (see **Figure 2** above) based on expected load demand and supply bids from power generators. RTOs commit to purchase power from generators based on day-ahead market clearing prices. Approximately 95% of energy sales within the RTO are transacted in the day-ahead market, with the remaining 5% transacted in the real-time market.[12] Real-time market clearing prices, also known as balancing energy prices, represent energy prices that reflect deviations of forecasted versus actual load and generation. By their nature, real-time prices are generally much more volatile than day-ahead prices. Typically, wind-influenced negative prices are reported based on real-time market outcomes. However, real-time power prices can impact both day-ahead clearing prices and bilateral power purchase agreements, as they can influence pricing expectations for contract negotiations.

---

[12] "Energy Primer: A Handbook of Energy Market Basics," Federal Energy Regulatory Commission, July 2012.

## Bilateral Contracts

Generators operating in RTO/ISO power markets can enter into bilateral power purchase contracts directly with load serving entities (LSE) that purchase power for distribution to end-use customers. These bilateral contracts can serve to establish a consistent price for electricity supplied by a power generating facility. From the power generator's perspective, the risk of low wholesale power prices is reduced and these contracts can provide a stable revenue stream for the generator. The mechanism for settling the power purchase price defined in the bilateral contract is sometimes referred to as contracts for differences. The difference between the market clearing price and the power purchase price in a bilateral contract is generally settled between the contract counterparties. While these types of contracts can mitigate the impact of low and negative wholesale power prices, since many bilateral contracts are on a short-term basis (e.g., one to three years), persistent low and negative wholesale prices could impact a generator's ability to negotiate future electricity purchase contracts.

## Uplift/Make-Whole Payments

Many RTOs use uplift/make-whole payments as a means to encourage power generation facilities to offer electricity at the marginal cost of production and at the direction of system operators. Uplift/make-whole payments are essentially compensation paid to some generators in the event that revenue received from the wholesale power market is less than marginal operating costs. While these payments do not guarantee dispatch, a price for electricity generated, or a threshold rate of return, uplift/make-whole payments, in essence, do provide some degree of assurance that generators might not operate at a loss.

## Cost Recovery and Amortization

In order to obtain sources of financial capital necessary to build a new power generating facility, each plant must have a revenue, operations and maintenance, and profitability plan that allows for all installation costs to be recovered over an asset's lifetime. Typically, capital cost recovery might occur within the first 20 to 30 years of plant operations. A project developer looking to construct a generating facility that might be located in a power market would need some degree of assurance that revenue from the combination of energy sales, capacity markets, and ancillary services, over time, would be adequate to pay for all necessary costs (i.e., capital, operations, maintenance, fuel, and finance) and provide an acceptable financial return. Alternatively, new power generating projects could potentially enter into bilateral contracts with LSE counterparties as a means of ensuring an adequate revenue stream that would attract investment capital. However, some older generators (generally more than 20 to 30 years old) may have achieved full cost recovery. Owners of these facilities may be more comfortable operating in wholesale power markets as they only need to recover fuel and operations/maintenance expenses. For these fully recovered facilities, wholesale power markets could be more lucrative than for power plants still amortizing and recovering capital costs.

## Negative Prices

Negative wholesale electricity prices can occur in RTO-managed competitive markets due to a combination of transmission congestion and a sufficient number of negative priced bids from power generators. In an RTO area where low-priced bid generators cannot deliver some of their production outside the area due to transmission congestion, the market model will lower the output of, or not dispatch, some generators based on their price bids. To ensure that they continue to be dispatched, generators may lower their bids even to below zero if necessary. If all the generators taken in a particular area had negative bids, the resulting LMP for the location will be negative. Negative LMPs are a disincentive to generate power at certain times and under certain conditions.

Some generators are willing to accept negative prices (essentially, the generator pays the RTO to provide power) for a variety of reasons, some of which might include: (1) maintenance and fuel costs associated with power plant shutdown and start-up may exceed the cost of the temporary negative price event, (2) technical difficulties with cycling nuclear plants on and off provide an inherent incentive to operate continuously, and (3) incentives are provided for electricity production, such as the production tax credit (PTC) that is currently available for wind power projects.

# Multiple Factors Can Affect Wholesale Electricity Prices

Wholesale electric power prices are influenced by a number of different variables. Some of the most common variables include: (1) economic activity and load demand, (2) weather, (3) fuel prices, (4) transmission congestion, and (5) zero marginal cost generation. As described in **Figure 2** above, wholesale prices are bid by generators based on the marginal (i.e., fuel) cost of production. Wholesale prices are cleared at the intersection of supply bid and load demand. As such, the amount of economic activity (load demand) and the price of fuel can significantly influence wholesale market clearing prices. For example, the MISO 2011 state-of-the-market report states that wholesale market prices were lower in 2011, compared to 2010, as a result of lower natural gas prices and lower loads.[13] In many RTO-managed markets, natural gas generation typically sets the market clearing price during peak demand times when wholesale prices are highest and when generator margin opportunities are the greatest. Low natural gas prices result in lower natural gas generation costs, which reduce the wholesale clearing price and therefore the revenue and margins available to all generators. Additionally, weather conditions directly influence load demand, and can move clearing prices up or down depending on the seasonality or abnormality of annual weather conditions. Also, as briefly discussed above, transmission congestion can cause clearing prices to fluctuate in certain locations within an RTO operating area. Finally, as will be discussed in further detail below, the addition of zero marginal cost generation can potentially reduce market clearing prices by shifting out the power supply curve, thereby reducing the market clearing price at a specified demand level during certain times throughout the day.

Since 2008, wholesale power prices in all RTO markets have declined. In some RTO markets, power price declines have been substantial (see **Figure 3**).

---

[13] "2011 State of the Market Report for the MISO Electricity Markets," Potomac Economics, June 2012.

**Figure 3. Comparison of All-In Wholesale Power Prices Across Multiple Markets**

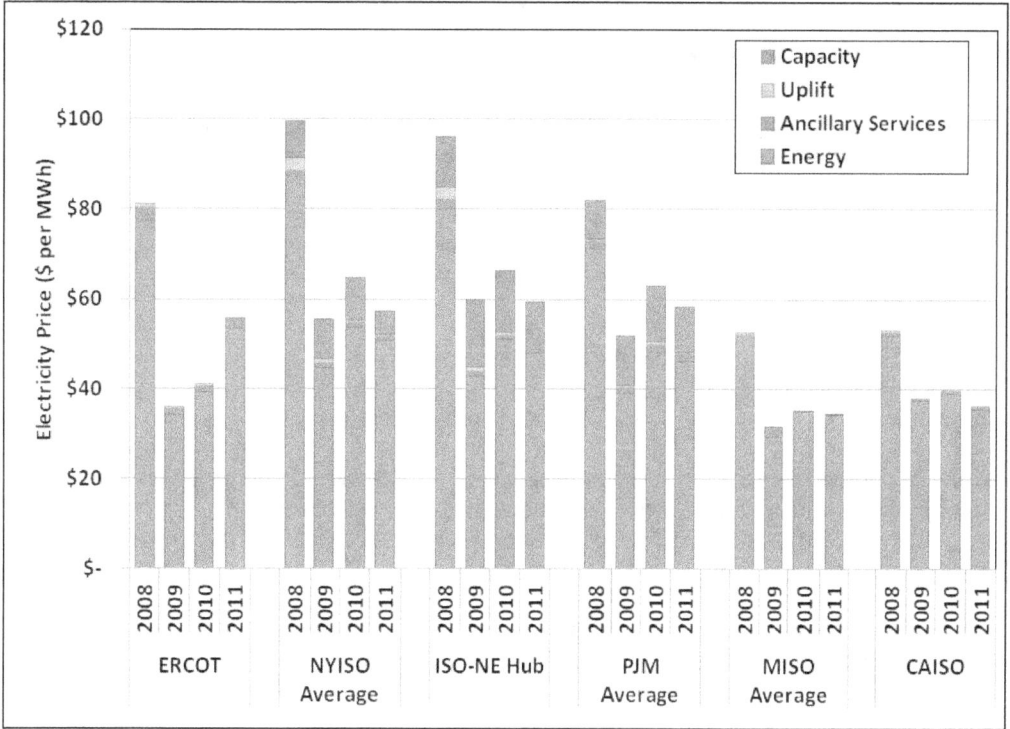

**Source:** "2011 State of the Market Report for the ERCOT Wholesale Electricity Markets," Potomac Economics, July 2012.

**Notes:** Per the source document: "the figure reports the average cost (per MWh of load) for energy, ancillary services (reserves and regulation), capacity markets (if applicable), and uplift for economically out-of-merit resources." The term "out-of-merit" refers to resources that have a marginal operating cost that is higher than the market clearing price during a certain period of time.

As indicated in **Figure 3**, wholesale power prices and price declines vary across each RTO. Arguably, the wholesale price declines observed since 2008 are primarily due to lower natural gas generation costs and lower electricity demand. The power price fall coincides with the 2008 financial crises and the associated reduction in electricity demand due to lower levels of economic activity.

Also, increasing U.S. natural gas production through the expanded use of hydraulic fracturing and horizontal drilling techniques contributed to low natural gas prices. Low natural gas prices translate into low marginal costs for natural gas electricity generation. Since natural gas generation is typically used to supply power during peak demand periods, market clearing prices were generally lower during peak demand times when power generators might otherwise be able to capture higher revenues and operating margins.

# Impacts of Wind Power in Competitive U.S. Electricity Markets

Various studies and reports, as referenced below, have been published regarding the wholesale electricity price impacts from the increased penetration of wind power. Considering the key concepts and definitions described above, the following sections examine three fundamental questions related to the impact of wind power in competitive electricity markets: (1) How might wind power affect market clearing prices?, (2) Does wind power contribute to negative wholesale electricity price events?, and (3) Does wind power impact electric system reliability?

## How Might Wind Power Affect Market Clearing Prices?

The addition of wind power capacity within competitive power markets can, in some markets and locations and under certain conditions, put downward pressure on electricity market clearing prices. Proponents of wind power generation argue that lower market clearing prices will reduce power prices for all consumers; therefore additional wind power can reduce the cost of electricity to a wide range of consumer groups. Opponents, however, argue that federal tax credits and state renewable portfolio standard policies result in lower market clearing prices, distort wholesale power markets, displace generation from base load power facilities, and unfairly impact the compensation received by existing generation assets. Both views are defendable to some degree; however each position assumes that energy sales are the only source of revenue and compensation available to all electric power assets. As an example, a theoretical "all-wind" power scenario would have a constant market clearing price at or near zero ($0.00), since the marginal cost (i.e., fuel) of generating a MWh of wind electricity is available at no cost. This, of course, is not economically feasible, since wind projects must capture revenue to pay for capital, operations and maintenance, and finance costs. In reality, electricity markets are inherently complex and can include multiple revenue opportunities. Nevertheless, the impact magnitude of lower market clearing prices resulting from increased wind power production and the economic impact to other power generators is uncertain and can be influenced by factors such as market rules, transmission congestion, and contractual conditions (e.g., bilateral contracts). In the 2011 PJM State of the Market report, the PJM market monitor states the following:

> Output from wind turbines displaces output from other generation types. This displacement affects the output of marginal units in PJM. The magnitude and type of effect on marginal unit output will depend on the level of the wind turbine output, its location, the time of the output and its duration.[14]

As illustrated in **Figure 2** above, electricity supply bids are provided based on the marginal (i.e., fuel) cost of generation, with wind power being the least expensive (at or near zero), since the fuel cost for wind generation is zero.[15] Nuclear power is typically the next lowest priced source of electricity as fuel costs for generation are minimal. As generation becomes more fuel intensive—for power generators that use coal and natural gas fuel sources—the marginal price begins to increase based on the cost of the fuel used and the efficiency of the respective power plants. At

---

[14] "State of the Market Report for PJM," Monitoring Analytics, LLC, March 15, 2012.

[15] Many wind power projects establish contracts outside of the RTO wholesale power market with entities to purchase electricity and/or renewable energy credits (RECs) on a long-term basis.

the intersection of electricity supply and power demand is what is known as the wholesale electricity clearing price (see **Figure 2**). Electricity is dispatched from each generator with bids at or below the electricity clearing price. However, all generators dispatched are compensated for their electricity at the clearing price level. For example, if a nuclear power plant bid $5.00 per megawatt-hour (MWh) and the clearing price was $20.00 per MWh, the nuclear generator would receive $20.00 for each MWh provided during the dispatch time period. **Figure 4** illustrates how additional wind power within a competitive market might theoretically impact the electricity clearing price for all generators.

**Figure 4. Illustration of Potential Wind Power Effects on Wholesale Electricity Prices**

(Hypothetical illustrative example for a one-hour interval in an RTO/ISO market:
not representative of actual RTO market clearing or dispatch results)

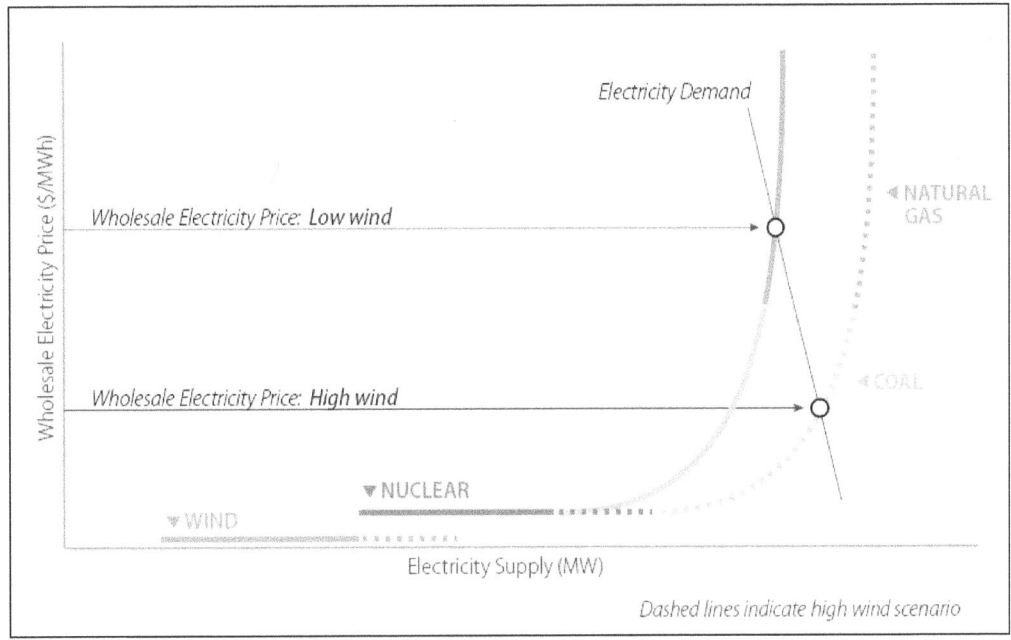

**Source:** CRS adaptation of "Wind Energy and Electricity Prices: Exploring the Merit Order Effect," European Wind Energy Association, April 2010.

**Notes:** Actual values for electricity prices and electricity supply are not indicated in this figure as they will vary by market, generation mix, time-of-day, and location. The "electricity demand" line is set at an angle to reflect additional demand that would likely occur as price levels decrease. However, electricity demand is somewhat inelastic and the slope of the line is steep in order to reflect the relative inelasticity of power demand.

**Figure 4** includes two electricity dispatch curves: one for a low wind scenario (solid lines) and one for a high wind scenario (dashed lines). As indicated, additional wind power results in shifting the dispatch curve to the right, thereby reducing the electricity clearing price paid to all generators, including the wind electricity projects. In essence, the addition of more wind power reduces the value received by wind projects, and all other power suppliers, selling power into wholesale markets. Since wind power is typically dispatched first, due to its zero marginal cost bid, additional wind power could potentially result in less electricity provided from other fuel sources (nuclear, coal, and natural gas).

The illustration in **Figure 4** is a hypothetical example only and is not meant to indicate the impact magnitude of adding a certain amount of wind power generation. To put into context how wind power might be affecting wholesale power prices, **Figure 5** provides the actual economic supply and dispatch curves for ERCOT, PJM, and the California ISO (CAISO).

**Figure 5. Actual Economic Supply Stacks for Three RTOs**

(ERCOT, PJM, and CAISO)

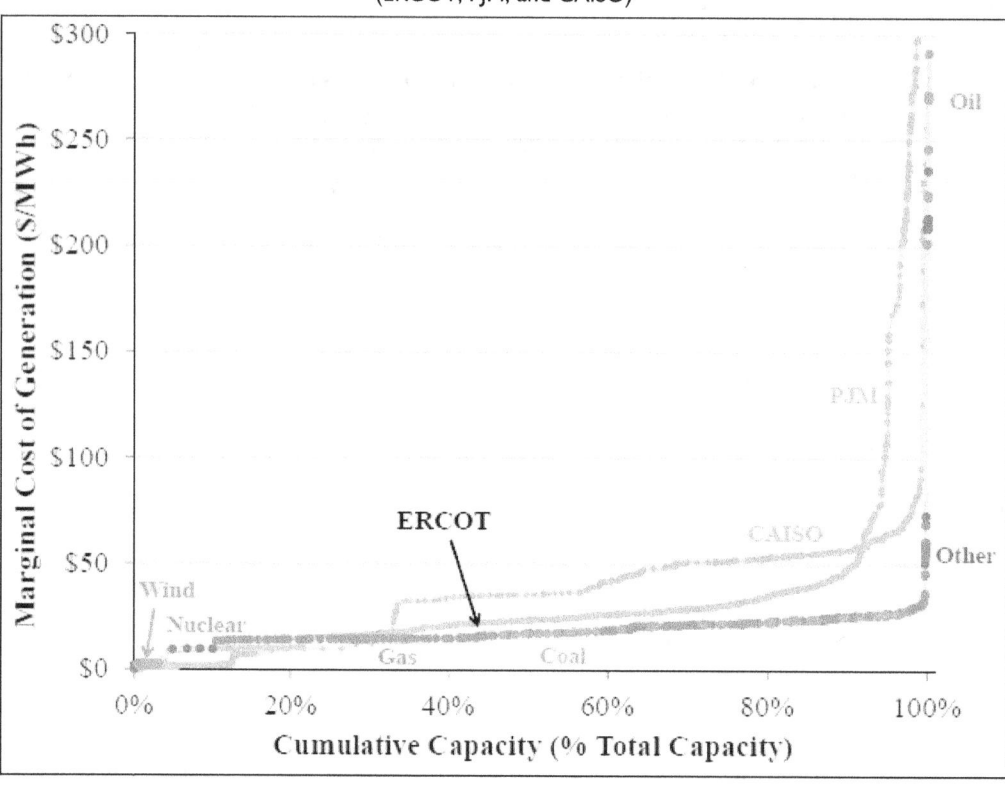

**Source:** S. Newall, et al., "ERCOT Investment Incentives and Resource Adequacy," The Brattle Group, June 1, 2012.

**Notes:** Figure notes as described in the source document: "Individual plants' marginal costs obtained from Ventyx (2012). To calculate plant marginal costs, Ventyx estimates VOM, fuel, and emissions prices. To calculate fuel costs, Ventyx estimates coal prices based on the last 3 months' delivered cost, natural gas prices based on 5/10/2012 spot prices via Intercontinental Exchange, and petroleum prices based on the 4/2012 ENERFAX price. Imports are not accounted for. Wind is derated to 20% of installed capacity."

In 2009, PJM conducted a study that considered the wholesale power price impacts of adding 15,000 MW of wind power in the PJM market. Results from the study indicated that the addition of wind power would decrease wholesale market prices by $4.50 per MWh. As a result, market-wide expenditures for wholesale power would go down.[16] For comparison, PJM's system-wide load-weighted average LMP was $45.19 in 2011.[17]

---

[16] "Potential Effects of Proposed Climate Change Policies on PJM's Energy Market," PJM, January 23, 2009.

[17] "State of the Market Report for PJM," Monitoring Analytics, LLC, March 15, 2012.

However, as discussed above, the wholesale clearing price may not reflect all of the compensation received by power generators. Electricity suppliers might have pre-negotiated power purchase agreements (PPAs) or other bilateral contracts with load serving entities that provide an agreed-upon power price regardless of the wholesale market clearing price. Additionally, power generators can also contract with LSEs to meet capacity needs that satisfy reliability and capacity requirements. Furthermore, other market factors such as additional economic activity and its associated load demand, along with rising natural gas prices, could result in upward pressure on wholesale market clearing prices. This upward price pressure could potentially result in opportunities for power generators to earn additional revenue and margins. Finally, exactly how much lower wholesale prices will impact other generators will be generator-specific and will include considerations such as location, contractual conditions, and the amount of cost recovery realized by each generating facility.

## Does Wind Power Generation Contribute to Negative Wholesale Electricity Price Events?

Several published reports, as referenced in the text below, provide insight into the relationship between wind power and negative wholesale electricity prices. Generally, reports reviewed for this analysis indicate that in some locations there is a positive correlation between wind electricity generation and the occurrence of negative real-time electricity prices. As discussed above, real-time prices are naturally much more volatile than day-ahead prices but the majority of market transactions are based on day-ahead clearing prices. In some markets, and during some time intervals, wind is setting the clearing price at a negative value by bidding into the market at a negative price level. The ability of wind to bid negatively priced electricity is a result of value received from federal production tax credit incentives and the potential opportunity to sell renewable energy credits (RECs) to third parties. Typically, transactions for these additional revenue sources occur outside of the RTO market operation.

Each generator typically operates under a different set of contractual conditions and transmission constraints; therefore the absolute impact of negative wholesale prices can vary on a case-by-case basis. In addition, the overall magnitude of the impact of negative prices on other power generators' revenue and margins is also uncertain and the lack of available data makes it difficult to quantitatively assess the direct impact of negative prices. Reasons for this uncertainty include many of the concepts and definitions discussed above: bilateral contracts—how much generation is contracted with LSEs at an agreed upon purchase price; uplift/make-whole payments—how much of the negatively priced real-time power is recovered through make-whole payments; and other revenue sources—how much revenue is captured through capacity, ancillary services, and other markets within each RTO. Furthermore, power generation revenues and margins are influenced by a number of factors such as fuel prices, weather, and electricity demand.

The time-of-day aspect is also an important parameter when considering negative prices. Wind power generation is greatest during nighttime and off-peak demand periods when wholesale prices are typically at their lowest levels. However, wind power generation is generally the lowest during peak demand periods when wholesale prices are typically at their highest levels. Negative power prices associated with wind power might generally occur at night when wind is producing at high levels. Large amounts of wind power generation can potentially contribute to transmission congestion and result in negatively priced wholesale power in certain locations. However, this negative price event is occurring when generator revenues and margins are typically at their lowest levels. During periods of peak demand, and high prices, wind typically generates power at lower levels, thereby allowing other power suppliers to take advantage of high-price/revenue/margin periods throughout each operating day. Nevertheless, since wind power generation typically peaks during night time (i.e., low demand) hours, the potential exists for traditional base load generators in certain locations to not get dispatched since they might be the marginal cost supply units.

As illustrated in **Table 1**, each RTO is unique with respect to its total generation mix and capacity, wind capacity, market types, and dispatch rules. Therefore, the impact of negative price events can impact generators in different ways depending on the RTO in which the generator operates. The following sections discuss available information about negative prices in MISO, PJM, and ERCOT.

## MISO

While the MISO wholesale electricity market has experienced negative price events in certain locations throughout the year, negative wholesale prices rarely occur and they are concentrated in certain hubs within the MISO service territory. **Figure 6** provides price duration curves for four different MISO hubs.

**Figure 6. MISO Real-Time Energy Price Duration Curve**

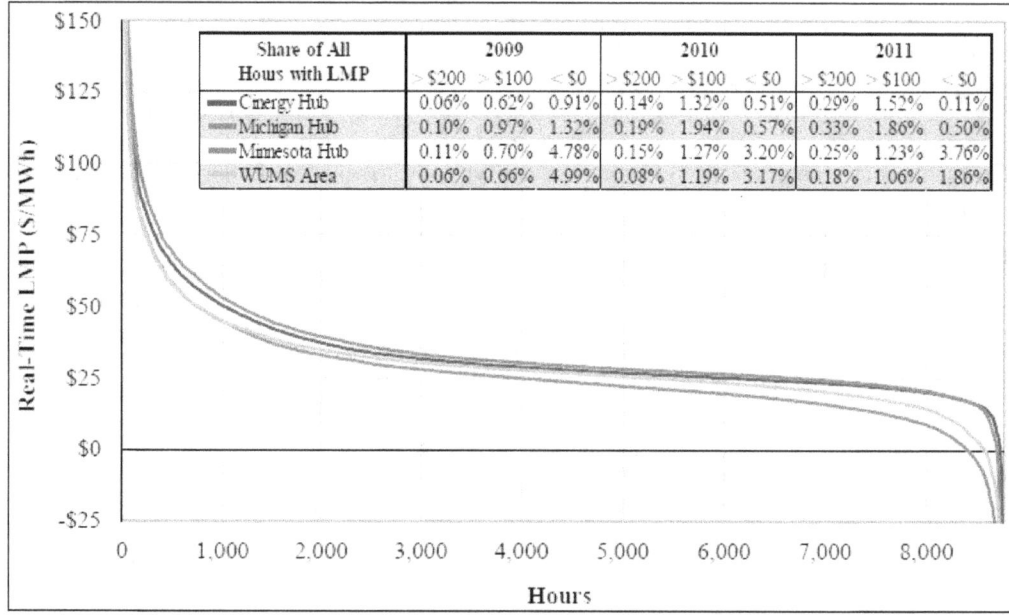

| Share of All Hours with LMP | 2009 | | | 2010 | | | 2011 | | |
|---|---|---|---|---|---|---|---|---|---|
| | > $200 | > $100 | < $0 | > $200 | > $100 | < $0 | > $200 | > $100 | < $0 |
| Cinergy Hub | 0.06% | 0.62% | 0.91% | 0.14% | 1.32% | 0.51% | 0.29% | 1.52% | 0.11% |
| Michigan Hub | 0.10% | 0.97% | 1.32% | 0.19% | 1.94% | 0.57% | 0.33% | 1.86% | 0.50% |
| Minnesota Hub | 0.11% | 0.70% | 4.78% | 0.15% | 1.27% | 3.20% | 0.25% | 1.23% | 3.76% |
| WUMS Area | 0.06% | 0.66% | 4.99% | 0.08% | 1.19% | 3.17% | 0.18% | 1.06% | 1.86% |

**Source:** "2011 State of the Market Report for the MISO Electricity Markets," Potomac Economics, June 2012.

**Notes:** Lines represent 2011 real-time LMP prices for each hub. The horizontal axis (Hours) represents the total number of hours in one calendar year (8,760). The purpose of the chart is to show, generally, the real-time hourly LMPs for four hubs in the MISO market over a one-year period. As indicated in the table located inside the above chart, MISO LMPs were above $100 for a few hours in 2011 and were below zero for a few hours. Although not specifically referenced in the figure table, the majority of LMPs (between 95% and 98% for 2011) ranged between $0 and $100.

Price setting in MISO by wind resources was enabled through the introduction of the Dispatachable Intermittent Resources (DIR) program in 2011.[18] Under the DIR program, wind production forecasts are used to determine maximum wind output for specific time intervals. Based on these forecasts, wind generation can be dispatched either up or down depending on system and economic conditions. As a result, the DIR program allows MISO more control over the variable nature of wind power generation, thereby reducing the need for electric system regulation. According to MISO's Independent Market Monitor (IMM):

> DIRs are wind resources that are physically capable of responding to dispatch instructions (from zero to a forecasted maximum) and can therefore set the real-time energy price. DIRs are treated comparable to other dispatchable generation, and therefore are eligible for all uplift payments and are subject to all requisite operating requirements. By June 2013, most wind units in MISO will be DIRs.[19]

MISO's IMM reports that in 2011 wind power generation set the wholesale price of electricity during certain time periods and in certain locations, at an average price of negative $20 per MWh.[20] The IMM attributes this negatively set wind price to the availability of federal production tax credit incentives.[21] However, negative price offers may also be incentivized by the opportunity of wind power projects to sell renewable energy credits (RECs) to entities in order to comply with state RPS policies.

The MISO IMM observes that *average* real-time energy prices in 2011, for the entire MISO market area, declined 1.9% compared to 2010. Two primary reasons for this decline, according to the IMM, were: (1) lower natural gas prices, and (2) lower average load demand. However, when adjusting for fuel costs, the IMM indicates that roughly two-thirds of energy price changes were due to a combination of declining load demand, increased net imports, and increased generation from intermittent resources (i.e., wind). Exactly how much wind generation directly contributed to the average price decline is not indicated in the IMM report. In 2011, wind power supplied 5.2% of electricity generation in the MISO RTO market.[22]

## PJM

In June 2009, PJM modified its market rules to allow all generator units to submit negative price offers into PJM energy markets. According to PJM's independent market monitor (IMM), "wind and solar units were the only unit types to make negative offers" since the new rule was established.[23] The IMM reports that an average of 935.5 MW, out of approximately 5,300 MW, of wind resources were offered at a negative price to PJM's real-time market in 2011.[24] These negative price offers are directly attributed to federal production tax credit incentives and the opportunity for wind generators to sell renewable energy credits (RECs) for each MWh of generation. The market monitor states the following:

---

[18] Ibid.

[19] Ibid.

[20] "2011 State of the Market Report for the MISO Electricity Markets," Potomac Economics, June 2012.

[21] Ibid.

[22] Ibid.

[23] "State of the Market Report for PJM," Monitoring Analytics, LLC, March 15, 2012.

[24] Ibid.

The out-of-market payments in the form of RECs and federal production tax credits mean these units have an incentive to generate MWh until the negative LMP is equal to the credit received for each MWh adjusted for any marginal costs. These subsidies affect the offer behavior of these resources in PJM markets.[25]

During calendar year 2011, wind represented 2% of the marginal generation used for the real-time energy market.[26] Wind was never a marginal generation source for the day-ahead market in 2011.[27] As discussed above, the day-ahead and real-time markets typically include 95% and 5% of energy transactions, respectively. In 2011, wind power accounted for 1.5% of electricity generation in PJM.[28]

## ERCOT

ERCOT is an often-cited example of how wind power generation can directly cause negative wholesale electricity prices in RTO-managed power markets. ERCOT, which includes the majority of Texas, has four primary pricing zones within its system (Houston, North, South, and West). With more than 10,600 MW of wind power installed,[29] out of a total installed capacity of approximately 84,000 MW,[30] Texas has more wind power capacity than any other state. The West Zone includes the majority of installed wind capacity in the ERCOT system, and this zone typically experiences the highest degree of price volatility in the real-time energy markets. **Figure 7** provides real-time wholesale price duration data for the four ERCOT zones.

---

[25] Ibid.

[26] Ibid.

[27] Ibid.

[28] Ibid.

[29] American Wind Energy Association website, http://www.awea.org/learnabout/industry_stats/index.cfm, accessed September 27, 2012.

[30] "2011 State of the Market Report for the ERCOT Wholesale Electricity Markets," Potomac Economics, July 2012.

**Figure 7. ERCOT Real-Time Price Durations Curves**

(Calendar Year 2011, in hours)

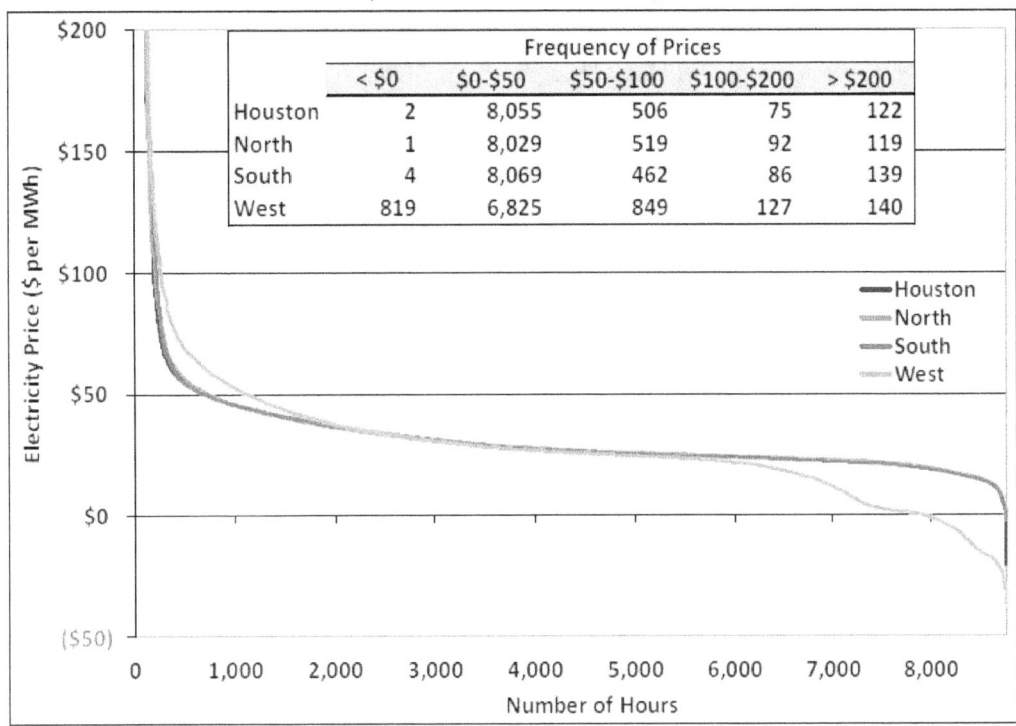

| | Frequency of Prices | | | | |
|---|---|---|---|---|---|
| | < $0 | $0-$50 | $50-$100 | $100-$200 | > $200 |
| Houston | 2 | 8,055 | 506 | 75 | 122 |
| North | 1 | 8,029 | 519 | 92 | 119 |
| South | 4 | 8,069 | 462 | 86 | 139 |
| West | 819 | 6,825 | 849 | 127 | 140 |

**Source:** "2011 State of the Market Report for the ERCOT Wholesale Electricity Markets," Potomac Economics, July 2012.

**Notes:** "Frequency of Prices" indicates the number of hours during a calendar year that prices were in each range. Lines represent 2011 wholesale electricity prices for each zone. The horizontal axis (Number of Hours) represents the total number of hours in one calendar year (8,760). The purpose of the chart is to show, generally, the hourly electricity prices for four zones in the ERCOT market over a one-year period. As indicated in the table located inside the above chart, the majority of LMPs ranged between $0 and $50. In the ERCOT West Zone, there were negative prices during 819 hours in 2011, while no other ERCOT region experienced more than four hours of negative prices. The lowest price was approximately -$40 per MWh. On the other hand, the duration of "high" (those greater than $50/MWh) prices was greater throughout all regions of Texas. Houston, North, and South price duration curves overlap in the figure.

As indicated in **Figure 7**, the ERCOT West zone experienced the majority of real-time negative price events within the ERCOT system in 2011. According to one report, 90% or more of ERCOT wind capacity is located in the West zone.[31] It is important to note, however, that the West zone also experiences the greatest duration of high energy prices (see **Figure 7**), although the disparity with other zones is generally less than that for negatively priced energy.

The profile of wind generation is inversely correlated with the load demand profile in ERCOT. Much like other regions of the country, when load demand is high, wind production is low and vice versa. Especially during nighttime hours, wind generation can potentially exceed load

---

[31] "The Relationship Between Wind Generation and Balancing-Energy Market Prices in ERCOT: 2007–2009," National Renewable Energy Laboratory, November 2010.

demand in a particular zone and the transmission infrastructure may not have enough capacity to transfer the wind-generated electricity to other demand centers. To resolve this situation, ERCOT may need to ramp down generation in order to balance supply and demand. Generators can offer down-balancing-energy services, which basically indicate the price level at which a generator will stop producing electricity and will instead purchase at the ERCOT clearing price for power needed to satisfy its production obligations. These down-balancing-energy prices are typically positive (i.e., greater than zero). However, according to a National Renewable Energy Laboratory report, wind generators in ERCOT can submit negative price offers since they receive federal production tax credits for each MWh generated, and they can also sell renewable energy credits for electricity produced.[32]

The Public Utility Commission of Texas (PUCT) is actively working to establish transmission infrastructure that would transmit wind power from West Texas to other load centers within the state. The program through which this infrastructure is being deployed is known as the Competitive Renewable Energy Zone (CREZ) Transmission Program.[33] Once CREZ projects are complete, negative price frequency in ERCOT's West zone may be reduced as wind power is transmitted to other demand centers. However, generators in other ERCOT hubs may be concerned that additional wind power could potentially put downward pressure on their revenues and operating margins.

## Does Wind Power Impact Electric System Reliability?

Two specific aspects of electric system reliability are typically discussed in the context of how wind power might impact electricity markets. First, RTO system operators must be concerned with real-time system operations and the ability to constantly manage supply and demand. Large amounts of wind electricity generation can potentially result in operational reliability issues due to the variable and sometimes unpredictable nature of wind power. Second, some arguments have been made that since wind power can potentially reduce revenues and margins available to all generators, the economic signals necessary to build new capacity for resource adequacy requirements are being distorted and the long-term reliability of power systems is being jeopardized. Each of these issues is unique, and each is addressed separately in the following sections.

### Real-Time System Operations

Wind power generation is naturally variable and wind electricity production can change outside of the system operator's control. Therefore, large amounts of wind generation can potentially create operational issues associated with managing the variable output of wind power facilities. A primary operational challenge that system operators could encounter when managing a large wind power fleet is the constant balancing of electricity supply and demand when power output can potentially change rapidly based on wind conditions. As a result, certain types of reserve capacity will likely be needed in order to respond to wind output variations. There are two specific reserve types that may be needed to manage the variable nature of wind power production: (1) operating

---

[32] Ibid.

[33] For more information, see the Public Utility Commission of Texas CREZ Transmission Program Information Center website, available at http://www.texascrezprojects.com/.

reserves,[34] and (2) spinning reserves.[35] Generally, these types of reserves would enable the rapid (within minutes in some cases) ramping and load following capabilities needed to accommodate sudden changes in wind power output as well as frequency and regulation services to ensure that the electric power system functions within normal operating parameters.

The National Renewable Energy Laboratory (NREL) has studied and analyzed the challenges associated with large amounts of wind power and has published several reports on this topic including the "Eastern Wind Integration and Transmission Study (EWITS)," which analyzed the reliability and integration aspects of a 20% wind power scenario in the Eastern Interconnect footprint.[36] The EWITS report indicates that the operational challenges associated with large amounts of wind power can be addressed with an expansion of the transmission infrastructure.[37]

Some U.S. power system operators are currently experiencing operational challenges associated with wind generation and are taking steps to address and manage some of these issues. As briefly discussed above, MISO has instituted the Dispatchable Intermittent Renewables (DIR) program that aims to provide the system operator with more control over the variable nature of wind electricity generation. Other system operators have implemented various curtailment policies that require wind generators to stop producing electricity under certain power system operating conditions. Additionally, improved wind forecasting and more frequent electric power scheduling and dispatch can potentially serve to address operational issues associated with wind output variability.

## Resource Adequacy and Capacity Margins

RTO system operators are typically concerned with the long-term operational reliability of the electric power system, and they generally monitor and plan for the power generation resources needed to satisfy expected load demand during a certain period of time (sometimes referred to as resource adequacy). A key metric used by RTOs to measure and track reliability and resource adequacy is "reserve margin." Reserve margins basically indicate the amount of excess power generation capacity available to the RTO when compared to the expected peak load demand for a given year. For example, if 2012 peak load is expected to be 1,000 MW and an RTO had a target capacity margin of 15%, then the RTO would need access to 1,150 MW of generating capacity in order to meet the target capacity margin.

Some groups may argue that by reducing market clearing prices (**Figure 4**) and contributing to negative locational marginal prices, wind power reduces the economic signals (revenue and profit) necessary to stimulate investment in new power generation facilities. The argument

---

[34] The National Renewable Energy Laboratory defines "operating reserves" as follows: "That capability above firm system demand required to provide for regulation, load forecasting error, forced and scheduled equipment outages, and local area protection. This type of reserve consists of both generation synchronized to the grid and generation that can be synchronized and made capable of serving load within a specified period of time."

[35] The National Renewable Energy Laboratory defines "spinning reserves" as follows: "The portion of operating reserve consisting of (1) generation synchronized to the system and fully available to serve load within the disturbance recovery period that follows a contingency event; or (2) load fully removable from the system within the disturbance recovery period after a contingency event."

[36] For more information about this study, see "Eastern Wind Integration and Transmission Study," prepared by EnerNex Corporation for the National Renewable Energy Laboratory, February 2011, available at http://www.nrel.gov/docs/fy11osti/47078.pdf.

[37] Ibid.

continues that this lack of investment might cause reserve margins to further decline, thereby putting system reliability at risk. As discussed in additional detail above, it is important to understand that wholesale electricity prices, typically the largest source of revenue available to power generators, are influenced by a number of factors (i.e., demand, weather, and fuel prices). Furthermore, the market design and operation of each respective RTO can also influence reserve margin and system reliability outcomes.

As indicated in **Table 1**, MISO, PJM, and ERCOT each have different market designs and different means to address resource adequacy and reserve margin issues. Although fundamentally very different, both MISO and PJM operate a capacity market that is designed to provide a source of income to generators that may not sell enough electricity to be economically viable, but are necessary to RTOs in order to satisfy target reserve margins and ensure system reliability. As more variable energy sources are added to an RTO system, the premium for reserve capacity could rise. If this occurs, generators providing capacity to meet resource adequacy requirements may see more revenue coming from capacity payments and less revenue from energy sales. As a result, the revenue profile for power generators may change. Ideally, net revenues would be adequate to incentivize generators to provide the capacity needed to maintain reserve margins. ERCOT, however, only operates an energy/ancillary services market.

In order to estimate and plan for resource adequacy and reserve margins, RTOs assign capacity credits (expressed as a percentage of nameplate capacity) to generators based on a generator's ability to supply needed power during peak demand periods. Capacity credits generally take into account planned and unplanned outages for power generators during a calendar year. Wind generators are typically assigned relatively low capacity credits to reflect the variable nature and generation profile of wind power. In the MISO RTO, wind generators receive a 14.9% capacity credit,[38] while PJM assigns wind generators a 13% capacity credit.[39] As an example, a 100 MW wind project in PJM would only contribute 13 MW towards resource adequacy and reserve margins. By comparison, a coal or nuclear generator may be assigned 80% to 85% capacity credits.

The 2011 MISO State of the Market report indicates that an inverse relationship exists between the amount of wind generation and capacity margins.[40] This relationship is associated with the capacity credits applied to wind projects and the reduced amount of wind power production during peak periods. As wind generation increases, capacity margins may decrease. As capacity margins go down the value of capacity should go up and be reflected in capacity market transactions. In theory, these capacity premiums would provide the economic signals necessary to stimulate additional capacity needed to satisfy target reserve margins.

Furthermore, power system reliability and resource adequacy is yet another complex element of RTO market operations that is influenced by multiple factors. One 2012 study assessed resource adequacy in the ERCOT market and the economic signals required to incentivize adequate generation capacity and reserve margins.[41] Unlike PJM and MISO, ERCOT only has an energy market. The study indicates that market conditions such as low natural gas prices, ERCOT's generation mix, and wind penetration have created challenging economic conditions for

---

[38] "2011 State of the Market Report for the MISO Electricity Markets," Potomac Economics, June 2012.

[39] "State of the Market Report for PJM," Monitoring Analytics, LLC, March 15, 2012.

[40] "2011 State of the Market Report for the MISO Electricity Markets," Potomac Economics, June 2012.

[41] S. Newall, et al., "ERCOT Investment Incentives and Resource Adequacy," The Brattle Group, June 1, 2012.

motivating new capacity additions.[42] Additionally, the study indicates that several environmental regulations may also present challenges to ERCOT's resource adequacy. Regulations evaluated include (1) Cross-State Air Pollution Rule, (2) Mercury and Air Toxics Standards, (3) Clean Water Act, Section 316(b), and (4) Coal Combustion Residuals Disposal Regulations.[43] However, CRS analysis indicates that environmental regulations may not be significant factors that affect system reliability.[44] Nevertheless, the combination of low power prices, low fuel prices, low levels of electricity demand, environmental regulation compliance, and the addition of zero marginal cost generation could potentially contribute to resource adequacy issues in RTO-managed markets.

# Policy Discussion

The immediate policy issue related to the market and economic effects of wind power is the future of production tax credit (PTC) incentives for wind projects. As briefly discussed above, the PTC has supported U.S. wind industry growth since being introduced in 1992. Under current law, PTC incentives will no longer be available to new wind projects on January 1, 2013. Debate about PTC incentives for wind power includes many aspects and considerations, the financial and economic impacts discussed in this report being only one. Proponents of extending the PTC incentive may contend that the wind industry supports thousands of U.S. manufacturing and project development jobs, supports many environmental objectives, helps diversify U.S. energy supply, and can potentially reduce consumer electricity bills. Opponents of a PTC extension may argue that wind is a mature electric power technology that no longer needs federal subsidies, the cost of the PTC is too expensive relative to the small amount of overall generation provided by wind, and the intermittent and variable nature of wind power may result in power system operational reliability challenges. Others have advocated a gradual phase-out of the PTC incentive as a way to eventually eliminate them while allowing the industry time to adjust to incentive reductions.[45]

The Senate Finance Committee reported the Family and Business Tax Cut Certainty Act of 2012 (S. 3521) on August 28, 2012. Among a number of tax-related provisions and extensions, S. 3521 includes language that extends the availability of production tax credits for wind facilities until January 1, 2014. The bill also modifies the definition of a qualified facility by allowing projects that start construction, rather than be placed in service, by January 1, 2014 to qualify for PTC incentives. This modification could be viewed as important to wind projects as it alleviates investment and development pressures that might result from having to place new projects into service by the end of 2013.

---

[42] Ibid.

[43] Ibid.

[44] For additional information see, CRS Report R41914, *EPA's Regulation of Coal-Fired Power: Is a "Train Wreck" Coming?*, by James E. McCarthy and Claudia Copeland. For additional information see, CRS Report R42144, *EPA's Utility MACT: Will the Lights Go Out?*, by James E. McCarthy.

[45] For additional policy discussion about the wind PTC, see CRS Report R42576, *U.S. Renewable Electricity: How Does the Production Tax Credit (PTC) Impact Wind Markets?*, by Phillip Brown.

# Concluding Remarks

Increasing amounts of wind power can potentially impact wholesale power prices in RTO-managed markets by possibly reducing market clearing prices and contributing to negative price events in certain locations during certain seasons and times of day. However, the absolute financial impacts of wind power generation are unclear due to the complex nature of wholesale power markets and the many variables that can impact wholesale electricity prices and generator revenues (i.e., location, natural gas prices, generation mix, and electricity demand).

Independent market monitor reports, as referenced in the text above, for the three selected RTOs (MISO, PJM, and ERCOT) indicate that wind power can contribute to negative price events; however, negative prices are more likely to occur at night when wind power generation is high, load demand is low, and electricity prices are low. However, during peak demand, when power prices are high, wind power generation is typically low. Therefore, wind power may have less of a price impact during times when generators can capture high revenues and earn high margins.

Finally, some studies suggest that wind power could potentially influence reliability issues or contribute to resource inadequacy.[46] However, each RTO has created a unique power market that is designed to incentivize reliability and resource adequacy within its operating territory. Nevertheless, should wind power continue to experience growth, it is uncertain whether current RTO market designs would function to ensure availability of the types of generation that would be necessary to both maintain resource adequacy and manage the variable and intermittent nature of wind power.

# Author Contact Information

Phillip Brown
Specialist in Energy Policy
pbrown@crs.loc.gov, 7-7386

---

[46] For additional information see S. Newall et al., "ERCOT Investment Incentives and Resource Adequacy," The Brattle Group, June 1, 2012; and F. Huntowski, A. Patterson, and M. Schnitzer, "Negative Electricity Prices and the Production Tax Credit," The Northbridge Group, September 14, 2012.

---